C0-CEP-271

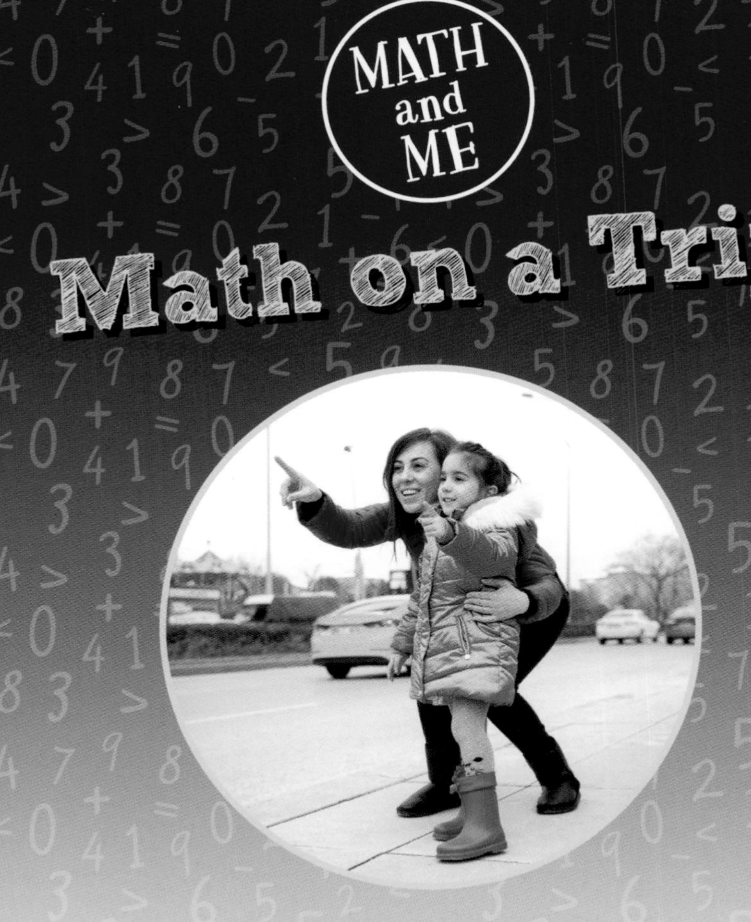

Math on a Trip

by Joanne Mattern

Red Chair Press Egremont, Massachusetts

Look! Books are produced and published by Red Chair Press:
Red Chair Press LLC PO Box 333 South Egremont, MA 01258-0333

Publisher's Cataloging-In-Publication Data
Names: Mattern, Joanne, 1963- author.
Title: Math on a trip / by Joanne Mattern.

Description: Egremont, Massachusetts : Red Chair Press, [2022] | Series: LOOK! books : Math and Me | Interest age level: 005-008. | Includes index and suggested resources for further reading. | Summary: "Math isn't just something you learn in school. It's an important part of the world around you. There are so many ways to use math on a trip. From telling time to measuring distance, readers will learn how using math can make travel more interesting"--Provided by publisher.

Identifiers: ISBN 9781643711331 (hardcover) | ISBN 9781643711393 (softcover) | ISBN 9781643711454 (ePDF) | ISBN 9781643711515 (ePub 3 S&L) | ISBN 9781643711577 (ePub 3 TR) | ISBN 9781643711638 (Kindle)

Subjects: LCSH: Mathematics--Juvenile literature. | Travel--Mathematics--Juvenile literature. | CYAC: Mathematics. | Travel--Mathematics.

Classification: LCC QA40.5 .M386 2022 (print) | LCC QA40.5 (ebook) | DDC 510--dc23

Library of Congress Control Number: 2021945371

Photo credits: Cover, p. 12, 15, 18, 19, 22, 24: Shutterstock ;p.1, 3–7, 9, 10, 11, 13, 14, 16, 17, 21, 23: iStock

Printed in United States of America
0422 1P CGF22

Table of Contents

On a Trip

Do you like to go on a trip? Taking a trip can be fun. You use math a lot when you **travel**.

How Far?

You are going to visit friends. It is a long trip. The map shows how far it is to get there.

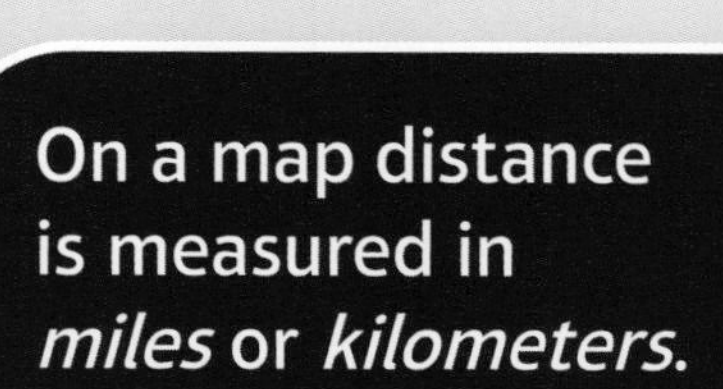

On a map distance is measured in *miles* or *kilometers.*

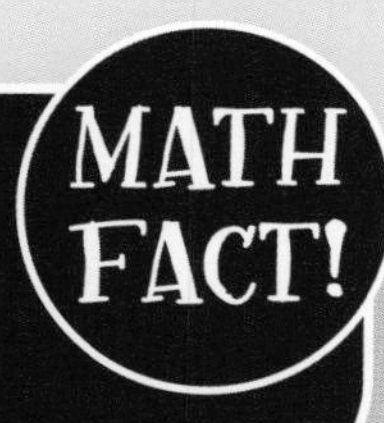

Your trip is 180 miles long. You can drive 60 miles an hour. How many hours will it take to get to your friends' house?

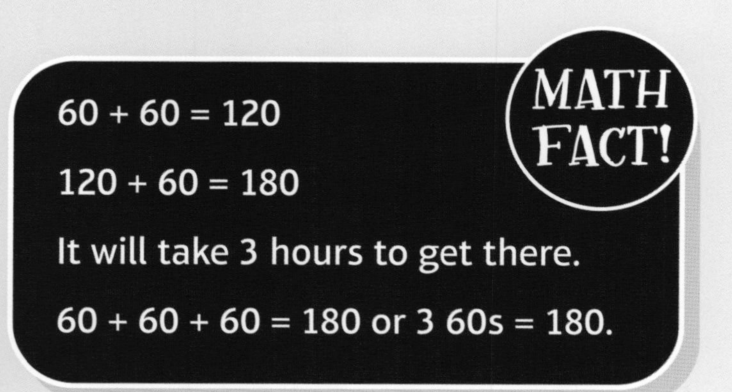

On the Road

You leave at 10:00 A.M. You drive at 60 miles per hour. Now it is 12:00. How far have you traveled?

60 + 60 = 120

After 2 hours, you have traveled about 120 miles.

MATH FACT!

Time for Lunch

Everyone is hungry! Let's stop for lunch. Your lunch costs $8. You give the **cashier** two $5 bills. How much is that? How much change will you get back?

MATH FACT!

$5 + $5 = $10

$10 – $8 = $2

You will get $2 back.

Counting the Way

You count road signs to pass the time. You count 10 blue signs. Your mom counts 12 green signs. How many signs did you count all together?

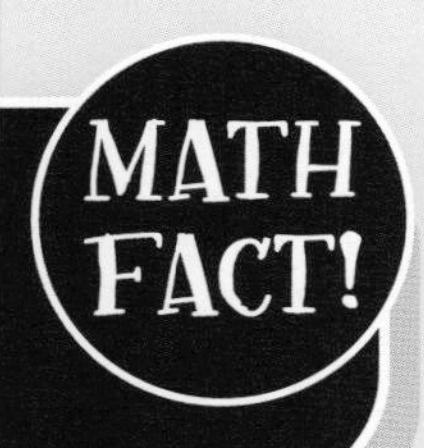

10 + 12 = 22

You counted 22 signs.

EXIT 55A
2
Santa Monica Blvd
EXIT ONLY
THRU
TRAFFIC
MERGE
LEFT

1
2
3
4
5
6
7
8
9
10

There are lots of trucks on the road! How many do you see? Let's count them.

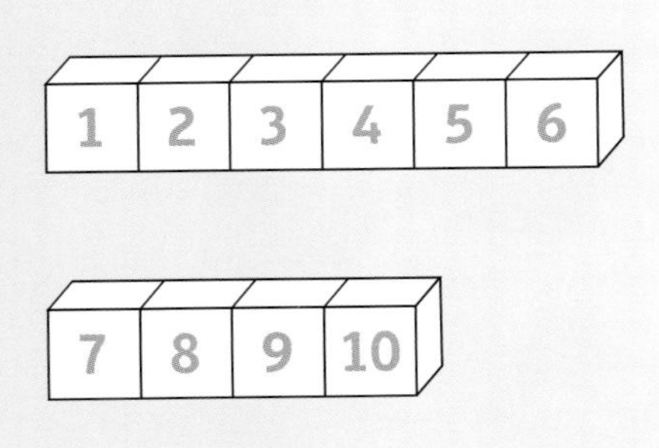

Almost There!

You stop to get gas. You buy 10 **gallons** of gas. Each gallon costs $3. How much money do you spend?

MATH FACT!

$3 × 10 = $30

You spend $30.

You say this multiplication sentence as 3 dollars times 10 gallons equals 30 dollars.

Here at Last!

We are at our friends' house! It feels good to get out of the car. Your friends come outside to meet you. How many of them are there? Let's count.

1, 2, 3, 4.

The family has
two adults and
two children.

A Fun Trip

You used a lot of math on your trip. How will you use math on the trip back home?

Words to Know

cashier: a person who takes money when you buy something in a store or restaurant

gallons: liquid measure equal to 4 quarts

travel: to go from one place to another

Learn More at the Library

Check out these books to learn more.

First, Rachel. *Count It! Fun with Counting and Comparing.* Abdo, 2016.

Mattern, Joanne. *I Use Math on a Trip.* Gareth Stevens, 2006.

Index

About the Author

Joanne Mattern is the author of many books for children. She loves writing about sports, animals, and interesting people. Mattern lives in New York State with her family.